# ROBOTS AND ROBOTICS

# MARGARET BALDWIN & GARY PACK

# ROBOTS
# AND
# ROBOTICS

A Computer-Awareness
First Book

Franklin Watts
New York | London | Toronto
Sydney | 1984

We would like to gratefully acknowledge the help
of the General Motors Corporation in sending us
information about their industrial robots.


Artwork by Jerry Gregory


Photographs courtesy of
Movie Star News: opposite p. 1, pp. 5, 7;
Twentieth Century Fox: p. 9;
New York Public Library Picture Collection: p. 22;
Unimation, Inc.: p. 28; General Electric Co.: p. 30;
NASA: p. 36; Naval Ocean Systems Center,
San Diego, CA: p. 37; RB Robot Corp.: p. 41 (left);
United Press International: p. 41 (right).

Library of Congress Cataloging in Publication Data

Baldwin, Margaret.
Robots and robotics.

(Computer-awareness first book)
Bibliography: p. 57-58
Includes index.
Summary: Describes robotics—the science of building
robots—and gives advice on careers in that field.
Also defines robots and describes their function in
industry, outer space, and the home.
1. Robotics—Juvenile literature. 2. Robots—
Juvenile literature. [1. Robotics. 2. Robots]
I. Pack, Gary. II. Title. III. Series.
TJ211.B34 1983          629.8'92          83-17070
ISBN 0-531-04705-9

7-84 Pub 8.90

# CONTENTS

*"I can't seem to remember owning a 'droid . . ."*

*Obi-Wan Ben Kenobi,*
*® Star Wars*

# ROBOTS AND ROBOTICS

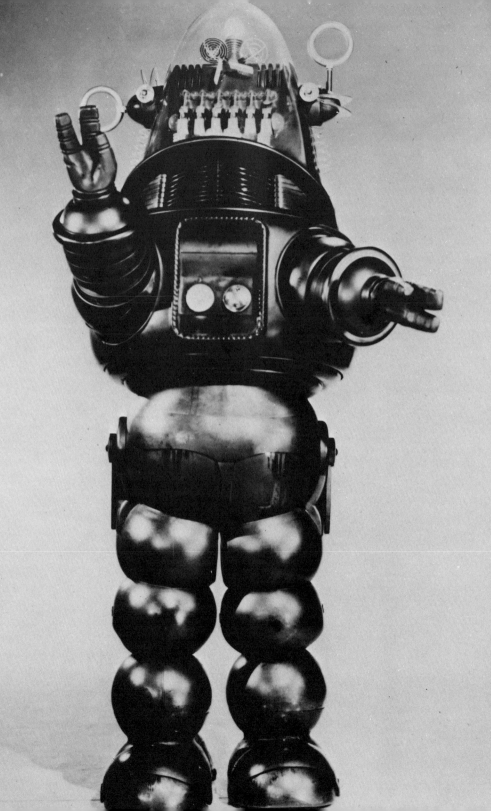

# PREFACE

One problem we faced when writing this book was defining the term *robot*. That may come as a surprise, since robots are part of our pop culture. But C-3PO, known as Threepio, would turn up his metallic nose at the robots currently working in our society today—if indeed they can be called robots. Some scientists insist that a true robot has not been developed yet! We found United States industry has a different view of *robotics*—the science of building robots—than Japanese industry, while the toy industry holds yet another view: any nonliving, mechanical device that can move under what *appears* to be its own power can be termed a robot. The metal creature with flashing lights who wanders around the shopping center shaking hands is commonly called a "robot." Actually it is operated by remote control and is no closer to being a robot than a battery-operated race car.

The robots we will be discussing are the closest things to true robots that have been developed. We urge you to pay close attention to the several definitions of robots you will find in this book. This will be very important in your understanding of robotics.

# CHAPTER ONE

# TALKING HEADS
# AND
# FRANKENSTEIN

*"Thou art a creature*
*of the magicians . . ."*

*the* Talmud

Robots have clanked their way through human dreams for centuries. Do you find that hard to believe? Perhaps you thought robots were a product of modern science-fiction writers.

No, the idea of robots has been with humans for a long, long time.

What is a robot? A robot is a machine with the ability to gather, study, digest, and respond to information. The last two functions are done independent of human control.

Another word that is used frequently in the science of robotics is *android*. This word is often used to mean the same thing as a robot. A shortened form of this word is " 'droid." "I'm only a third-degree 'droid and not very knowledgeable about such things as transatomic physics," remarks one of the most famous robots ever created—Threepio from George Lucas's *Star Wars*.

Actually, Threepio is not a " 'droid" at all, according to the precise definitions of "robot" and "android." An android is a creature put together from actual human parts or parts made to look human. A robot is a machine created out of mechanical parts.

Therefore, Threepio is—technically—a robot. The most famous android came from the dream of a teenage girl, Mary Shelley, who wrote the story *Frankenstein* in the early 1800s. Many science-fiction books and movies use the two terms, android and robot, to mean the same thing. In this book, however, we will follow the precise definitions.

Accounts of robots are found as early as ancient Greek mythology. According to legend, the god Vulcan, who was skilled in all things mechanical, created servants made of pure gold to wait upon him. He also made tripods—three-legged stools—with golden wheels that would roll in and out of rooms at his command.

There are many stories from ancient Egypt about robots. The Egyptians believed that statues had souls and if a statue was asked a question it would answer. Several early writings include accounts of wonderful Egyptian statues that spoke and even moved their arms. Archaeologists—scientists who study ancient civilizations—have discovered some of these statues, which are hollow. A priest would put his head into the statue's head and talk through the mouth, making people believe that the statue was really speaking. There were strings attached to the arms so that the priest could cause the statue's arms to move.

Stories of robots were very popular in the Middle Ages. Sir Francis Bacon, a writer and scientist who lived in the sixteenth century, was supposed to have had a talking head made out of bronze. According to legend, Bacon wanted to build a wall of bronze around the entire island of Great Britain to protect the people from attack by their enemies. The talking head was sup-

**Frankenstein's monster was an android, composed of various parts of dead human beings.**

posed to tell how this could be done. But when the head was finished, it would not speak. Bacon called upon a spirit who gave the head the power of speech, but the spirit warned that anything the head said must be answered immediately or it would never speak again.

Two priests came and sat before the head day and night, waiting for it to talk. At last they grew so tired they fell asleep. A servant was instructed to wake them the instant the head spoke. Of course, no sooner had the priests fallen asleep than the bronze head suddenly opened its mouth and said, "Time is." The servant didn't think a statement like this was important enough to wake his masters, so he let them sleep. An hour later the bronze head said, "Time was." Once again, the servant didn't consider this comment important, and he let the priests sleep. Finally, the bronze head snapped, "Time is past," and that was the last anyone ever heard from it.

The story of the *golem*, which also comes from the Middle Ages, tells of a man-made creation that got free from the control of its maker. There are many legends about the golem. According to one, a rabbi (Jewish priest) made the golem out of clay and dust to protect the Jewish people from attacks by their enemies. In a solemn ritual the golem was brought to life when the name of God was placed in its mouth. The golem wandered through the streets and saw the suffering of the Jewish people. It was so affected, that, instead of protecting people, it began to kill everyone it considered an enemy to the Jews. This horrified the rabbi, who found he could not stop his creation, and he was forced to destroy the golem.

**This scene from the 1917 film *The Golem* shows the humanlike clay monster who was brought to life.**

[6]

In part, the story of the golem inspired Mary Shelley to write about a doctor who created a man out of the parts of the dead. Dr. Frankenstein's creation is not really evil—there is a touching scene in which the monster listens to an old man playing the violin—but like the golem, the monster becomes violent and attacks its creator. The authors of these stories may have been trying to say that we were never meant to meddle with certain things, such as creating intelligent life artificially.

The science-fiction theme of robots and androids who kill their creators is very popular. Perhaps this is because humans have always been frightened of the machines they have created. In the early days of railroads, many people were terrified of the huge mechanical monsters that roared through the countryside, and many predicted they would bring about the end of the world.

The word "robot" was first used in a play written by Karel Capek. It means "slave" in Czechoslovakian. The play was titled *R.U.R.*, which stands for Rossum's Universal Robots. Actually, Capek's robots were androids by definition. They were people created by scientists out of real human parts. But they lacked two characteristics that would have made them human—the ability to have children and the possession of a "soul."

"They've astonishing memories," explains one of the scientists. "If you were to read a twenty-volume encyclopedia to them, they'd repeat it all to you with absolute accuracy. But they never think of anything new." The robots in *R.U.R.* become so intelligent they destroy the human race, but they, too, are doomed to destruction because they will wear out. The last living scientist creates two robots and gives them souls, so there is hope the robots will be able to "go forth and multiply."

In the 1960s, when the computer age was leaping ahead with astonishing speed, machines were built that could "think" much more quickly and much better than people. There were many stories about the evil of machines that were too intelligent. Hal, the

**R2-D2, a famous robot, receives an important message from Princess Leah in a scene from *Star Wars*.**

computer in the book *2001*, murders almost the entire crew of a spaceship. Captain Kirk of the star-ship *Enterprise* fights a computer that is trying to take his job in the television series "Star Trek." This computer, too, kills innocent people.

Isaac Asimov, a famous science-fiction writer, calls this fear the "Frankenstein complex." To calm these fears, Asimov created the "Three Laws of Robotics." Robots, he says, should be programmed according to these laws:

1. A robot may not injure a human being, or through inaction allow a human being to come to harm.

2. A robot must obey the orders given it by human beings, except where such orders would conflict with the First Law.

3. A robot must protect its own existence as long as such protection doesn't conflict with the First or Second Law.

Asimov's character, Robbie the Robot, becomes the friend and playmate of a small child and even saves her life. R2-D2, known as Artoo, and Threepio from *Star Wars* also follow Asimov's laws. They care about the humans they serve and can even "feel" happiness, fear, and worry.

Are there robots in existence today? What do they do? How do robots "think" and act? How do robots move? Can robots be programmed to "feel"? Do we want to program robots to be able to "learn," or would such programming be dangerous?

We will explore these questions in the following chapters.

# CHAPTER TWO

# THE ROBOTIC
# "MIND"

*"I've been ordered to take you to the bridge.
Here I am, brain the size of a planet, and they
ask me to take you down to the bridge. Call that
job satisfaction? 'Cos I don't."*

Marvin the robot,
The Hitchhiker's Guide to the Galaxy

How does a robot "think"?

To answer this question, we are going to create our own robot. A robot such as the one we will be describing will be possible in the near future. It differs from complex robots such as Threepio who exist only in movies or in the distant future.

Like the human body, a robot's body requires a mind to control and direct its actions. The "mind" in a robot is a computer. We have already developed industrial robots with computerized "minds" that can perform tasks considered simple for humans but which, until recently, were considered too complicated for machines to perform.

These robots will be described in a later chapter. They are dependent on humans for programming their "brains."

If you are told to go across the room and pick up the cat, you do so without really thinking about what you are doing. You are using skills you have been learning ever since you were born. You must walk, for one thing. Your brain must tell your legs how to move. You see the room you are standing in and you know how far you have to walk to reach the cat. You see the cat sitting on the floor, and you know that it is a cat and not a dog or a plant. You know how to pick up the cat so that you will not hurt it. All of these actions that seem so simple to you would be very difficult for a robot.

When scientists first began to study how to program, or instruct, the computer "brains" of robots to obey commands, they tried to "copy" the human brain. These early attempts failed because the human brain is much too complex to be duplicated by electronics. In fact, there is still a great deal about the human brain that we do not understand. Scientists realized that it was not necessary or possible to copy the brain in order to create a "thinking" machine.

One of the most important functions of the human brain is decision making. We make countless decisions every day—from what we want for breakfast to where to go to college. Scientists know that one of the most important functions of a robot is the ability to make decisions, so they study how people make decisions. In one study, they observe how people make decisions while playing games such as poker, chess, or backgammon. These games all involve clearly defined goals. People must make decisions to reach these goals.

A robot can be given a memory. A robot can be programmed to obey. A robot can even be programmed to make choices, basing its decisions on information stored in its memory. The difficulty facing scientists is to create a "brain" that will allow the robot to take in new information, store it in its memory, and act upon it when necessary. Robots who could do this could function independently of humans because they could *learn*.

In order to explain the complex process a robot's "mind" goes through in order to follow a simple instruction, let us imagine our robot already has the ability to understand human speech, obey commands, and learn. We are going to command our robot to go across the room and pick up the cat. Let's see what happens.

First, our robot must be able to understand us when we speak to it. Its brain is a computer, of course, but there is no keyboard on which to enter commands or programs. These programs, called *software*, must previously be written and then stored in the "brain." The robot's software must be able to "hear" normal human speech and translate it into computer language—special language a computer understands. Languages currently being developed for this purpose are LISP and LOGO.

Let us assume we have programmed our robot with a vocabulary and a list of commands. The robot recognizes that it is being given a command. It takes our words, turns them into computer language, compares them with words stored in its memory, turns them back into language we can understand, and repeats them back to us so we know the robot understands the command. If the robot did not understand, it would ask for further information before proceeding.

The robot's "brain" must break that one simple command into parts of speech in order to understand precisely what it is being told to do. It breaks the sentence into subject, verb, and object. Here's how the robot "thinks" over the command "You go across the room and you pick up the cat."

*You:* subject—meaning I, the robot

*go:* verb active—meaning to travel

*across the room:* first object of command—where I am supposed to go

*and:* connecting word—there is more to the command coming

*pick up:* verb active—meaning to lift

*the cat:* object—small, furry animal, second object of the command, what I pick up

We must speak our command in language that is clear and concise so that we do not confuse our robot.

Our robot repeats the command, indicating it understands us. Next, the robot must find the object to be picked up. This it does by scanning the room with its *optical sensors* or "eyes." We will explain how these work in chapter 3 on the mechanics of a robot.

When we hear the word "cat," we automatically call up an image of a cat. The image comes from our memory—we remember what cats look like because we have seen them before. Therefore, if there are two objects across the room and one is a cat and the other is a rosebush, we know from memory which one is the cat. If we had never seen a cat before, but knew it was a small, furry animal, we could use that information to tell the difference between the two objects. If we had no idea which was the cat and we brought back the rosebush, we would be told that we were wrong. We would learn which one was the cat and would remember the next time.

Here is one possible way for the robot to perform this task. The robot scans the area across from it and sees two objects—one is a cat, the other is a rosebush. It must decide which of these objects it has been commanded to pick up. The robot compares the two objects with a three-dimensional model of a cat it has stored in its memory, considering such characteristics as size and shape. It must also be able to match the picture in its memory with the real cat no matter what position the cat is in—lying down, standing up, or sitting. This process of comparison lets the robot know immediately which object is the cat.

Having identified the cat, the robot must now "go across the room." It has to decide which is the most direct path to the cat. First it must scan the area between itself and the cat. Are there any objects in its way? If there are, it must choose a path that is not blocked. The brain commands the body to move. The robot will then start to travel across the room.

**This diagram illustrates the path the robot must take to avoid any obstacles and still reach the cat.**

The robot has reached the cat. (We will assume that the cat is used to the robot and will not run away as it approaches!) It will have to pick up the cat. This task seems very easy to us. We know from experience that we should not pick up a cat by the tail. But the robot doesn't know this unless (1) we have programmed the robot to know how to pick up a cat, or (2) the robot "learns" how to pick up a cat by experience—which might be hard on the cat!

The robot does not understand "pain." We know that picking up a cat by its tail hurts the cat. We can imagine the pain the cat is feeling because we have felt pain ourselves. But a robot does not feel pain and has no way of imagining pain. Telling the robot to pick up the cat without hurting it has no meaning for the robot.

[15]

We do not have time to program every command into our robot; we want it to "learn." Let's say it understands the command "be gentle." The robot has pressure sensors in its "hand," or manipulator, that tell it when it is putting enough pressure on the object to lift it. The pressure sensors are tuned to a point termed "gentle" in the robot's memory. This may be so gentle, however, that the cat falls out of the robot's "hand." We then command the robot to grasp the cat a little tighter. This time it squeezes the cat, and the cat yowls in pain. We may, at this point, have to actually put the robot's mechanical hand on the cat and demonstrate how to pick it up. This procedure would then be entered in the robot's memory, and in the future the robot would know how to pick up the cat.

Look at all the work and technology required to program a robot "brain" to do a task that we could perform in seconds! Let's now study the complex piece of machinery that is the robot's body and how it works with the computer "brain."

# CHAPTER THREE

# THE ROBOTIC
# BODY

*"Help!" Threepio yelled, suddenly frightened
at a new message from an internal sensor.
"I think something is melting. Free my left leg
—the trouble's near the pelvic servomotor."*

Star Wars

Our imaginary robot does not look much like a human. A robot does not have to look like a human, however, to be able to function efficiently. Hal, in the book *2001*, could be considered a robot; we could think of the entire spaceship as his body. Robots like R2-D2 resemble roving garbage cans. Industrial robots look like giant arms without bodies at all. Robots can be designed to perform many functions, not all of which require human shapes.

Our robot's body is made of heavy-duty, industrial plastic. We chose plastic instead of metal because it is lighter in weight. The robot stands about one meter (1 yd) high at its "shoulder." The cone-shaped head is about twenty centimeters (8 in) high. The body is cylindrical and moves on wheels.

We chose a three-wheel system for our robot. There are others, some of which include "legs" so that the robot can climb up and down stairs. Some of these systems have as many as six to eight "legs." We chose the three-wheel system because it is sim-

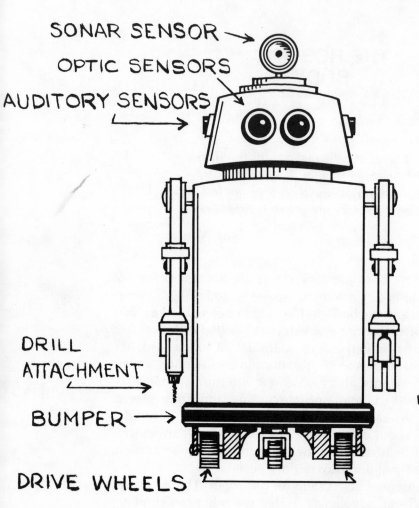

SONAR SENSOR →

OPTIC SENSORS

AUDITORY SENSORS →

DRILL
ATTACHMENT →

BUMPER →

DRIVE WHEELS

GREGORY

THUMB

FINGERS

DETAIL OF HAND

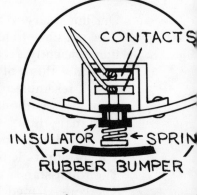

CONTACTS

INSULATOR → ← SPRIN

RUBBER BUMPER

CUT-AWAY DETAIL C
CONTACT SENSO

pler to build and control. The wheels, located on the bottom of the robot's body, are set in a triangular pattern. The two back wheels are the "drive" wheels, and they propel the robot either forward, backward, or sideways. Each drive wheel can be driven independently of the other, allowing the robot to turn right or left. The front wheel can move in any direction. The wheels are controlled by the robot's "brain"—the computer—through the use of *stepper motors.*

A stepper motor is an electronic motor that moves in steps according to instructions from the computer "brain." The motor divides one complete rotation of the wheel into one hundred steps. Each time it receives an electronic impulse from the "brain" it moves one step. If it received ten impulses it would move ten steps. These types of motors are used in the joints of the robot as well as the wheels. The "brain" knows the exact circumference of (distance around) each wheel so that it knows how many "steps" it will have to take in order to reach its goal. These motors are also called *servomotors.* A servomotor is an electronic motor that sends impulses back to the "brain," telling the computer how many "steps" it has taken.

When the robot is commanded to go straight ahead, the computer sends electronic impulses to the wheel motors. Both rear wheels turn at the same rate, moving forward. If the robot is ordered to turn right, the right wheel remains in the same position and the left wheel turns to the right. If the robot is commanded to back up, both wheels would move in reverse at the same speed. The robot's "brain" keeps track of the distance it has traveled by counting the rotations of the wheels.

The robot has a "bumper" with *contact sensors* around the

**The various parts of
our imaginary robot**

[19]

bottom of its body. A sensor sends a signal to the robot's "brain" when it "senses" something by touch and sight and sound. If the robot bumps into something, the contact sensor closes a switch which sends an impulse to the "brain" telling the robot it has run into an obstacle. The "brain" notes which sensor sent the message and determines what side of the robot the obstacle is on. It then tells the robot how to move in order to get around the object. The brain also remembers the location of the object so that the robot will not run into it again.

The body of the robot houses the computer "brain," the power source, and the skeleton of the robot. Power sources currently being used for robots are lead-acid batteries—which are just like the batteries used in cars—and nickel-cadmium batteries. There are drawbacks to both of these, however, and scientists are looking for better power sources. Lead-acid batteries are very heavy and bulky, and they also wear out quickly. Nickel-cadmium batteries are much lighter, but they are also bulky and very expensive. Some robots today are simply "plugged in" to an electric outlet. Future power sources being considered are fuel cells, similar to those used in the space shuttle. They generate electricity by a complex chemical reaction. These are a great deal like batteries, but they are much lighter and very efficient. Their major drawback is that they are very expensive.

Another excellent source of power might be atomic energy. An atomic battery would be small, light-weight, and long-lasting. The primary disadvantage in using atomic energy is that it would require heavy shielding to protect humans and delicate electronic parts.

Broadcast power is another energy source for robots, and this is what we have chosen for our robot. Let's say we have a power generator located in the basement of the house that sends out radio waves which the robot receives and converts into electricity. This is *not* remote control, since the robot is only receiving power, not instructions. The robot is still in control.

The "arms" or manipulators are the most complicated parts of the robot. The arms are joined at the body, allowing for movement up and down. There is another joint at the elbow which can bend up and down and also rotate in a complete circle. Our robot has a three-fingered "hand" with a "thumb" and two fingers. A thumb is very important for using tools. Some robots have hands that are more like claws without any joints. We have given our robots two fingers with one joint each so that they can bend and hold objects.

If you want to pick up a cat, you just extend your arm and pick it up. You see how close you are to the cat and, while you probably don't know how long your arm is, your brain tells you if you need to move closer in order to reach the cat. You do this without thinking about it.

The robot's brain knows just exactly how long the robot's arm is. It knows how each segment of the arm is positioned. It knows in what directions the joints can move and, since the joints are controlled by stepper motors, it can tell how far each joint has moved. By knowing all of this and using information from other sensors on the robot, the brain can position the arm to perform the tasks it is instructed to perform.

Pressure sensors in the robot's hand let the robot "feel" the object it is picking up and keep the robot from grasping it so tightly it would break or be injured. This is known as *negative feedback*. Feedback from the sensors tells the brain when the pressure is enough. This information has either been programmed into the computer or it has learned it on its own through trial and error, by picking up something and dropping it, picking up something and holding it, and picking up something and crushing it.

Our robot has only one three-fingered hand. The other "hand" is a special-purpose hand that is not like a human hand at all. We can attach an electronic drill to this arm or a welding torch or a can opener or even a mop for mopping floors.

The head of the robot contains the optic sensors, or "eyes,"

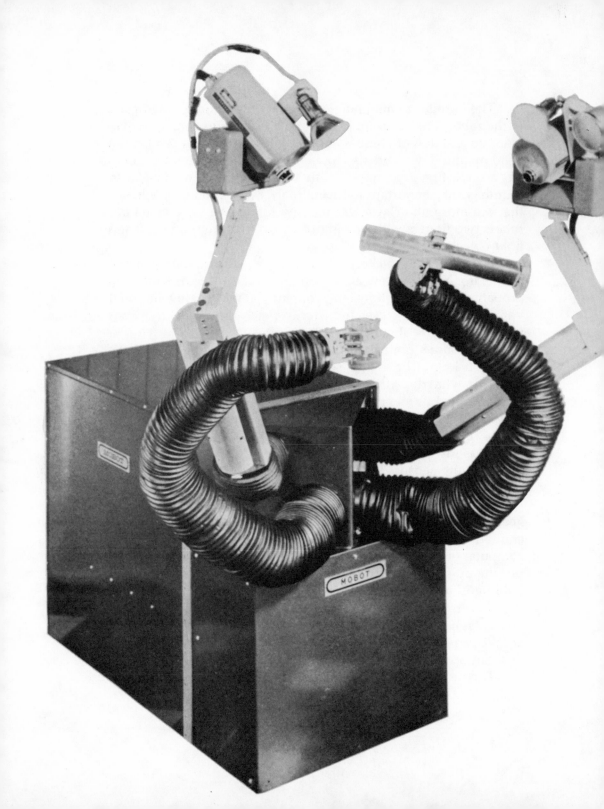

the auditory sensors, or "ears," and a sonar device. Of course, these could be located on other parts of the body—maybe the arms, for example—depending on what is most practical.

The robot's "eyes" are two cameras placed in the front of the head. These two cameras give our robot *depth perception*—the ability to see if one object is closer than another object. We are not born with depth perception. Babies do not have depth perception, which is why you will often see small children try to take hold of a flower that is printed on wallpaper. The child has not learned to tell the difference between a real flower and a picture of a flower. The child will learn this without even thinking about it. However, depth perception is proving very difficult for scientists to build into the systems of robots.

Robots are currently using video cameras for "eyes," but there are drawbacks. Video cameras are bulky, heavy, and require a high-voltage power source. Very bright lights can burn out the vacuum tubes in a video camera. In the future, scientists will probably use cameras that do not require vacuum tubes. These cameras will "see" an object, then turn the image of that object into electric signals that will feed into the robot's computer "brain." For a robot to pick up a cat, the "brain" must tell the robot which object is the cat and which is the rosebush. The "eyes" will also be used to "see" any objects that are in the robot's path and pass this information along to the "brain," which can then tell the robot how to avoid them.

The "ears" on our robot are *auditory sensors*—sensor units that pick up sounds. Just as humans have two ears, robots have two microphones, located on either side of the head. We want our

**The Mobot Mark II, designed by Hughes Aircraft, is another example of a robot with arms and sensors.**

[23]

robot to hear our vocal commands and know the source of the commands. Thus, if we say "bring the cat to me here," the robot knows where we are standing without necessarily seeing us.

The difficulty with robots "hearing" vocal commands is in turning the spoken language into electronic impulses that the computer "brain" can understand. As we have said before, when we talk to a computer "brain" we must be very precise and say exactly what we mean. If we tell our robot to "go through the door," that is exactly what it will do—go right through the door! We must tell the robot to "open the door and go through it."

The third function located on the robot's head is the *sonar device*. This device sends out sound waves that bounce off objects and return to the robot telling it the location, size, and shape of objects up ahead. Scientists have studied bats, which have such well-developed sonar systems that they can fly through a room filled with tiny wires without touching one of them. A bat sends out ultrasonic (beyond the range of human hearing) sound waves through the nose and mouth. These waves hit objects in the bat's flight path and return to the bat through its ears. The bat's brain "reads" these sonar waves, just as your brain "reads" the sounds you hear, and tells the bat where to fly. Robots could use the sonar waves to help judge distances as well. The sonar tells the robot what is in front of it. We could make sonar revolve, giving the robot information about what is behind it.

A robot's "brain," unlike the human brain, has a limit to what can be stored in it—although this limit may be very large. Three-pio was "familiar with six million languages." But, as we have stated, robots like the *Star Wars* robots are in the distant future.

We have learned how a robot would "think" about the command to pick up the cat. Here is how it would physically react to the command: "You go across the room and you pick up the cat."

The robot must turn its head and locate all surrounding objects with its "eyes" to see if any of them match the picture of a cat stored in its memory. The robot's head can turn a complete circle (360 degrees) so that it doesn't have to walk around to look at things. When the robot has located what its "brain" tells it is a cat, the sonar device sends out sound waves which bounce back off the cat and tell the "brain" how far away the cat is. The robot also uses the sonar to warn it of any objects in its path.

The sound waves return and tell the robot it is three meters, or 10 feet, from where the cat is sitting. Sonar also warns that there is a lamp between the robot and the cat. The robot's "eyes" also "see" the lamp. The brain tells the robot to move to the right one meter, or 1 yard ("x" number of wheel rotations), and then move forward. This way it will avoid the lamp.

Just as it starts off, however, the robot runs into a roller skate that has been left on the floor. The skate is so close to the base of the robot that the sonar and the "eyes" fail to detect it. The contact sensors on the bumpers around the bottom of the robot tell the brain the robot has struck an object. The brain knows from the position of the sensor that the roller skate is directly in front of the robot. It tells the robot to move over half a meter (20 in) and then return to the original program. This done, the robot rolls across the room, heading for the cat.

When the robot reaches the cat, it moves to within arm's length of the animal. The robot's brain knows how long the arm is and places the robot in the correct position. It now gives the robot precise information on extending the arm, reaching down, rotating the hand, and grasping hold of the cat. The robot has learned from us how to hold a cat so that it will not harm the animal. It reaches down and the fingers of the hand open. The fingers close gently around the cat. The pressure sensors on the fingers tell the brain when the robot is holding the cat tightly enough not to drop it, but not so tightly that the cat will be hurt. The arm raises and the robot stops, waiting for the next command.

As we have said, our robot exists only in the future. But robots are alive and well and working today. They are working in industry and hobby robots are "playing" around the house. Robotics is what's happening now!

[25]

# CHAPTER FOUR

## ROBOTS: ALIVE AND WELL AND WORKING TODAY

*"All the doors in this spaceship have a cheerful and sunny disposition. It is their pleasure to open for you, and their satisfaction to close again with the knowledge of a job well done."*

The Hitchhiker's Guide to the Galaxy

There are real robots in existence today, but they do not look or act like the robots we see in movies or read about in science-fiction books. These robots perform very important jobs for humans. They are called *industrial robots*.

What is an industrial robot? According to General Motors, a robot is "A reprogrammable, multi-axis, mechanical manipulator." *Reprogrammable* means that the machine can be programmed to do more than one task. *Multi-axis* means that the arms of the machine can move in a number of different directions. A *mechanical manipulator* is simply a machine that "handles" or manipulates an object. In the case of industrial robots, that can mean anything from spray-painting a car door to welding.

There are a lot of differences between industrial robots and Threepio. For one thing, industrial robots are in fixed places. If they move at all, they can move only on tracks that take them up and down the assembly line. These robots can't walk across a room. They have no bodies or heads. Some cannot "see" and would be just as likely to spray-paint *you* as they would a car body. They usually are just "mechanical arms" and a computer "brain." But they are capable of performing important jobs.

Robots are being used widely in the automobile industry because they are perfect for *assembly-line work*. In assembly-line work, a person stands near a moving belt or line that brings the objects to be put together. The worker generally performs just one or two tasks, then the part moves on and another part takes its place. In spray-painting a car door, for example, the line brings a car frame past, and the worker sprays paint on the frame. The painted frame moves on and another frame is brought up to be painted. At the General Motors plant in Doraville, Georgia, robots spray-paint cars as they come by on the assembly line. The robots are actually arms mounted on rails to allow them to move with the assembly line. There are seven joints where the arm can turn, move up and down, and side to side, so that it can paint in such areas as the inside of a car. Two robots spray-paint a car at the same time, working with a third robot arm that opens the car door after the outside of the car has been sprayed, allowing the robots to paint the inside. One computer "brain" controls the painting arms and the door-opening arm. The computer contains instructions for painting any of the many different body styles that might come down the assembly line.

The robot arms are "taught" to do their jobs by humans. They are trained by humans who grasp a handle attached to the robot arm and then guide it through the motions of the job, showing it how and where to spray the paint. These motions are "memorized" by the computer and later the arm performs the same motions without human help.

Spray-painting cars is a dangerous job. Paint fumes are poisonous if inhaled, and lead in paint is dangerous if it is absorbed through the skin. Human workers have to be protected from such dangers, but there is no need to insure the safety of robots. They

**These industrial robots perform dangerous welding operations on an automobile assembly line.**

also do the job better and more efficiently than humans. Robots spray the paint on evenly every time. They do not waste paint or get careless and bored doing the same job over and over.

Another dangerous job robots are performing for humans is *arc welding*. In arc welding, electric currents are used to melt two pieces of metal so that they join. The temperature of the metal becomes extremely hot and can severely burn human flesh. The high-voltage electric current could kill a human who comes into contact with the electrodes. The arc produces ultraviolet light that can burn the retina of the eye. Arc welding also is very dirty work. Robots can perform this dangerous, unpleasant job, protecting humans from injury.

Other uses for robots in American industry today include loading and unloading heavy material in dangerous areas. For example, robots load frozen slabs of meat in meat lockers where the temperature must be kept below freezing. Another type of operation currently being performed by robots is putting a block of steel into a machine tool and positioning it so that the machine can drill or grind it. Robots are also being used in forging operations, where hot metal must be accurately hammered into shape by repeated blows from a heavy instrument. Not only must the robot load and unload metal that is red-hot, but it must also be able to handle material that has changed shape during the forging operation.

Robots are being used in mold-making, in which metal or plastic is injected into dies or molds under high pressure. Here again, the robot works with material that is very hot and dangerous.

Robots are becoming more and more widely used in Japan. The Japanese hope to have completely automated plants in the future. In these plants, robots would do all of the work, directed by only a few human "supervisors." What would be some of the advantages? Energy costs would be lower. Robots can work in the dark. They do not need a well-lighted area like humans since they

**The General Electric GP132 industrial robot system
is used to lift heavy refrigerator cases.**

don't have to "see" what they are working on. They do not need heat in the winter or air conditioning in the summer. Robots increase productivity because they can perform jobs more quickly. In one Japanese plant, robots are being used to connect tiny, hair-thin wires to small computer chips. This job takes humans using a microscope three days because the wires are so fine they cannot be seen with the naked eye. A robot now performs the same job in ten minutes, greatly reducing the cost of the product.

Robotics in industry is just beginning. Scientists and engineers are currently trying to develop industrial robots with "senses." In particular, they want to develop a robot that can "see." It's not as simple as connecting a robot to a camera. As we found out working with our robot, the difficult part is in making the robot understand what it is "seeing."

The automotive industry wants robots that can choose a certain part from all different kinds of parts. The robots would have to be able to "see" the part, pick it up, and then attach it to the object it is assembling. The robot would also have to "see" if the part was broken or made incorrectly and then throw it away.

This is being done using *pattern recognition*. The robot we designed knew the difference between a cat and a rosebush because the image of a cat was programmed into its "brain." The industrial robot has the image of the part it needs programmed into its computer "brain." A camera attached to the computer is focused on the conveyer belt. Several different parts come into view. The camera takes a picture of all the parts and sends that picture to the computer. The computer can't "see" the different parts as we can. All it sees are various patches of light and darkness. The computer must recognize the correct pattern of light and darkness and thus "see" the outline of the part it is searching for. It compares this outline with the outline in its memory and decides if this is the part it needs.

The above process is very complicated. If two parts are touching, for example, there may be no clear, simple outline of light for the computer to find. If one part is sitting on top of another part, it is difficult for the computer to tell that these are two different parts. Problems like these will need to be solved before such robots can become operational.

Scientists are also working on developing robots that have the ability to "feel." If a robot could "feel" a part with its "hand" it would not need to be able to "see" it. This robot could reach into a parts bin and "feel" for the proper part. Sensors on the "fingers" would feed the shape of each part to the "brain," which would match the shapes with those stored in its memory. Robots are now being used to inspect parts to make certain that they are not defective.

There are many problems in the field of industrial robotics. Scientists and engineers are confident they can solve most of them in the near future. But a major concern is the threat of robots replacing humans in the work force. As we have seen, robots have been used primarily in jobs that are dangerous and dirty. In the future, however, robots will begin taking over jobs that people like and need since robots work better, faster, and cheaper than humans. People want to know what will happen when their jobs are taken over by machines. How will they support their families? Will they be retrained? Whose responsibility is it to retrain them? How will they live while they are being trained? Some people say that robots shouldn't be used because they will put people out of work. Others say that the growing robotics industry will eventually employ as many people as the computer and automobile industries do today.

Companies say that robots will benefit people a great deal because machine-produced products will cost less and be of better quality. Energy will be saved by using robots. These are difficult questions, and there are few answers right now.

One thing is certain, however, people are going to have to be well-educated in the future. If robots are going to be doing much of the work, there will be little need for unskilled workers.

# CHAPTER FIVE

## ROBOTS IN SPACE: FACT, NOT FICTION

*"We might even imagine a kind of Martian Johnny Appleseed, robot or human, roaming the frozen polar wastes in an endeavor that benefits only the generations of humans to come."*

Carl Sagan, Cosmos

Humans have always been curious about whether there is life on other planets. Scientists in the 1870s were able, through the use of telescopes, to detect what they believed were canals on Mars—a series of lines criss-crossing each other. This caused great excitement because it seemed to mean that there was intelligent life on Mars capable of building these "canals." But careful study indicated that canals filled with water could not exist on Mars because the temperature was too low and the atmosphere was too thin. What, then, were the "canals"? Many claimed they were simply optical illusions (such as a mirage).

With the space age, it has become possible for us to find out more about Mars and other planets. We sent men to the moon to bring back samples and to perform tests on the surface. However, there are many problems involved in sending humans to Mars and other planets. First, it would take many months to reach

Mars, the closest planet. Such a trip would require great quantities of food, water, air, and fuel. While we have the technological skills to overcome some of these problems, it would be very expensive and possibly dangerous to send humans to Mars.

There is another problem. Scientists believe that there might be life on Mars, though not intelligent life. There may be life in the form of tiny microorganisms such as bacteria and viruses. These could be much different from the small life forms on earth and could possibly be harmful to humans and animals. What we needed in order to study the surface of Mars was a craft that could perform human tasks.

At first scientists considered sending remote-controlled craft, machines controlled by people on the earth by radio signals— similar to radio-controlled model cars and airplanes. There were many drawbacks, however. It would take too long for the radio signals to travel from earth to Mars. For example, if you are on earth using remote control to drive a vehicle on Mars and you see that your vehicle is coming to the edge of a cliff, you would imme- diately step on the brake, telling the vehicle by radio signal to stop. By the time your signal reached the vehicle, however, it already would have tumbled over the cliff. Even if response time wasn't critical, directing the machine to perform tasks such as picking up rocks from the surface could take hours to complete.

Scientists turned to robot probes. The first probes that were sent out in the 1960s simply took photographs of the planets. These early probes were sent to Venus and Mars by both the United States and the Soviet Union. They used robots primarily for navigation—keeping the spacecraft heading in the right direc- tion—and for taking photographs and sending the data back to earth. These robots did not have a body or even "arms." They were like Hal in *2001*, an intelligent computer mind which con- trolled the operating functions of the probe—the entire ship was a robot.

One of the most sophisticated of these probes was the Viking

orbiter and lander. Viking was designed to photograph the surface of Mars with the orbiting portion of the spacecraft and to find a suitable landing site for the lander portion of the spacecraft. This information was sent back to earth, where a map of Mars's surface was made. Scientists studied this information and then fed the coordinates of what they considered an ideal landing site to the Viking. Once the command was given telling the Viking lander to descend to the surface of Mars, Viking took over and landed itself. Viking had a variety of tasks to perform under the control of the onboard computer. Although the computer had been programmed before it left earth, it could change its programming somewhat to deal with the variety of conditions it might encounter on Mars.

Viking performed many important tasks. It had a mechanical arm which reached out beyond the blast area surrounding the spacecraft and scooped up soil for analysis. This extending arm was necessary because scientists were afraid that the blast from landing the spacecraft might alter the surrounding soil, or kill any tiny life forms that might be present. The mechanical arm returned the soil sample to a test chamber on board. Viking then performed a series of experiments on the soil samples to find out what the soil was made of and if there was life in the soil. This data was sent back to earth. Viking also took photographs of the surface and sent these back to earth. Viking ran tests on the atmosphere of the planet. It also had seismometers on board which could register "Marsquakes." Viking collected all of this data and then, on command, transmitted it back to earth. Because Viking is a robot, it doesn't need any life support or feel lonely, so it could have collected data for years. It might still be doing so, but a faulty command was sent from earth that erased the information in its memory that told it in what direction to point its antenna to find earth.

The future of robots in outer space is every bit as exciting as science fiction. We are currently developing probes that will be

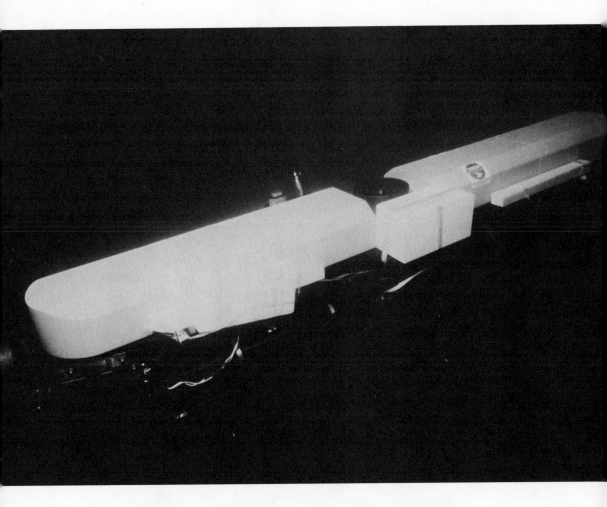

*Left:* the Viking lander, sent in 1976 to explore
the surface of Mars, is actually a robot programmed
to perform scientific experiments. *Above:* the Naval
Ocean Systems Center in San Diego is developing a
free-swimming submersible robot, designed to inspect
and monitor underwater cables and pipelines.

able to land on the surface of a planet, take samples, and return to the earth. Our first voyage to a star will be taken by a robot, because it will probably take thousands of years to travel from here to a star other than our sun.

Meanwhile, back on earth, robots are being used to explore a territory every bit as mysterious as outer space—the floor of the ocean. It is difficult for humans to explore the ocean floor because, just as in outer space, we need some form of life support. Another problem is the crushing pressure found miles below the surface of the water. And there are other problems—dangerous animal and plant life, the corrosive effects of salt water, powerful ocean currents, storms on the surface of the water, and the intense darkness. Robots are now being used to inspect underwater telephone cables, repair pipelines, and inspect and repair offshore oil-drilling platforms. Most of these are not free-moving robots, but those which depend upon humans for remote control. A future goal is to eliminate the need for human supervision.

The military has also become interested in robotics. In the future, military robots may be used for everything from issuing uniforms and equipment to detecting enemy minefields, digging ditches, and handling contaminated materials during nuclear, chemical, or biological warfare.

# CHAPTER SIX

## ROBOTS AROUND THE HOUSE

*"Robbie was constructed for one purpose
really—to be the companion of a little child.
His entire 'mentality' has been created for
the purpose. He just can't help being faithful
and loving and kind. He's a machine—made so.
That's more than you can say for humans."*

Robbie *by Isaac Asimov*

Just think how great it would be if you could tell your robot to make your bed, pick up your books, hang up your clothes, and have your room dusted by the time you get home from school!

Actually, this isn't such a wild dream. Several people, many of them teenagers, have built "hobby robots"—real working robots that operate around the house. This would include commercially manufactured robots such as Hero I.

As yet, these are really only interesting toys. They do little real work. But these robots and those who have built them will show us the way into a future when robots will be in every household.

One young person designed a robot that operated on musical tones—there were different tones for different commands. The robot had claws it could open and close, so it could pick up objects and move them around. It even contained a prerecorded tape so that it could "talk." The designer programmed the robot to travel five feet in one direction, turn, and continue 1.5 meters, or five feet, in another direction, while holding a vacuum sweeper in its claw!

Another young person designed a robot with sonar. It could enter a room and "learn" where objects were, either by bumping into them and detecting them with its sensors or by its sonar device. After the robot discovered where all the objects were located it would find the doorway and leave. Unfortunately this robot had a few "bugs." Sometimes the bumper switches would stick, making the robot think it was constantly running into something. The robot would dash across the room wildly, trying to get away!

Although these robots seem very primitive they are really giant steps forward. What is important about all of these robots is that they each operate independently. Many machines have been designed that can do very complex tasks such as serving people food on trays, shaking hands, and talking to people, but these machines are operated by remote control. Robots of the future will be able to "think" and act on their own.

Already many household computers are being programmed to help us operate our homes more efficiently. In some ways, these could be considered "robots"; the entire house is the robot's body. Although these robots cannot move around, they perform some interesting and useful functions. Home computers can be used to control our use of energy. They measure the temperature outside and inside, then turn on the heat or the air conditioning as needed.

**Two examples of domestic robots**

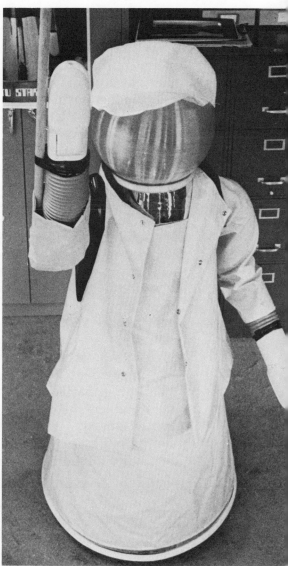

Home computers can also be used for protection. They can detect an intruder and automatically call the police. They then turn on all the lights and sound a siren to scare the intruder away. Computers can be programmed to communicate with smoke-sensing devices and call the fire department if fire is detected.

Some computers sense when a person enters a room and automatically switch on the lights, turning them off when the person leaves. They can be programmed to turn on the coffee pot in the morning so that the coffee is prepared when we wake up or turn on the oven to cook dinner while we are away. They answer the phone and take messages.

Of course, household computers cannot move. They cannot pick up after us or vacuum or mow the lawn. But as we have seen, with young people designing robots today, it will not be long before we have household robots that will provide us with much useful service as well as entertainment. In science fiction, robots like Asimov's "Robbie the Robot" and Ray Bradbury's Electronic Grandmother (actually an android because "she" was made of parts designed to look human) are used as babysitters. These robots, however, come to "care" for the children very much, and both robots risk their lives to save the children under their protection.

This is a problem that has long fascinated science-fiction writers and one that scientists are beginning to study in earnest. What will be the relationship between people and intelligent robots? Do we see them as Frankenstein monsters, capable of destroying us, or as lovable as Artoo?

# CHAPTER SEVEN

## COULD A
## ROBOT CRY?

*"I mean, where's the percentage in being
kind or helpful to a robot if it doesn't
have any gratitude circuits?"*

The Restaurant at the
End of the Universe

How will humans respond to robots in the future? This is a question we should be thinking about now. Unfortunately, it is a question often left to writers of science fiction.

Just as automobiles drastically changed our lives, so robots will have a profound effect on our future lives. The automobile made our lives easier, but it also created many problems. Cars allow us to travel faster and easier. We have become a more mobile society. But cars also produce air pollution and thousands of people die in automobile accidents every year. Of course, it is not the machine itself that causes these problems, but the people who build and operate them.

We have a very real, very basic fear of machines, even though we depend on them. We have created machines that can work much faster than we can. They are stronger than us. They never grow tired or get bored. But machines are blind and unfeeling. They can injure a human being without knowing they have done so and just keep right on working. In the future, it will be possible

[43]

to create robots that will think and learn as well as or better than humans. We also expect to develop robots that will have the ability to think and learn *on their own*. This frightens some people who fear intelligent robots will get out of control.

Humans, who are imperfect, fear that if we create a "perfect" robot or android we would not be able to stop it if it decided to take charge. Hal, in the book *2001*, becomes a "killer" robot. It is this fear which led to Asimov's "Three Laws of Robotics" presented in the first chapter. Could some evil person be stopped from building a "killer" robot by Asimov's "Three Laws"? If the technology is available, some people may use it for their own selfish purposes. Therefore, some people believe we should not try to develop intelligent, thinking robots.

What do you think? How would you solve this problem?

One solution popular in science-fiction novels is to program robots with human emotions. Threepio reacts to situations as a human would, by feeling fear, caring, or worrying. Robbie the Robot is programmed to "care" for children. We can, even now, program a robot to give the appearance of having human feelings. For example, we could program our robot to yell "ouch" if it ran into a table. We know the robot cannot feel pain, yet it could be programmed to act as though it did. Can Robbie—as a machine—really feel love and concern for a child? Or is it merely programmed to react as though it cared? Is there any difference between a machine acting as if it cares and real human caring?

What will happen if we program robots with human emotions? Fear is one of the strongest human emotions. It is an emotion essential to our survival. It is an emotion essential to animal survival. We are afraid of snakes because we know some snakes are poisonous. This fear keeps us from being bitten. A robot would need to be afraid of things in order to protect itself, but a robot wouldn't need to be frightened of a snake. It might, however, be dreadfully afraid of water since water could make it short-circuit. Should we program robots with emotions like fear? In

[44]

order for the robot to survive and keep functioning, it would need to be able to react to danger on its own, without a human having to watch it all the time. Yet fear could lead the robot to injure a human in order to protect itself. Should the robot have the right to protect itself?

Finally, what problems will arise when humans and robots live and work together? As we have seen, many jobs in business and industry are being performed now by *artificial intelligence*— robots with computer programming enabling them to "think." They are performing jobs people used to do, and performing those jobs better, faster, and cheaper. In many cases, however, this means people lose their jobs to robots.

We know that in the future an even larger portion of our lives will be dependent on robots. How will people get along with machines? How do you argue with a machine? Will people come to care for the machines they work with? It is human nature to give animals and machines human traits. Scientists already see this happening in plants where humans are working with robots. Is this good or bad?

Then, too, robots will give humans a lot more free time. If we have a robot do all the housework, we won't have to spend time cleaning our rooms, cooking dinner, or vacuuming the carpet. It sounds great, but how will we spend our free time? Will we use it creatively or waste it? Will we become bored?

Robots can be used for many good purposes, but will we take advantage of these opportunities? Will we be open to developing new skills and discarding ones that are no longer necessary? Since, for example, computers can do multiplication faster than we can, is it necessary for us to memorize the multiplication tables? Should we instead study the *uses* of mathematics instead of the mechanics?

It is important for you to think about these questions seriously, because you and your generation will be working to find the answers.

# CHAPTER EIGHT

# YOUR FUTURE
# IN ROBOTICS

*"Imagination is more important
than knowledge."*

Albert Einstein

If you are interested in a career in robotics, start planning now. There are subjects you can study in school to prepare for working with robots and androids. Other subjects you will need to study in upper grades and college. Some of these are fairly obvious, but a few you may find surprising!

The field of robotics is a complex one. It will probably not be possible for one person to possess all the knowledge needed to assemble complicated robots and androids. Undoubtedly, teams of people will work together, each sharing his or her own special skill. Teams will work on robot bodies and teams will work on robot "brains"—the computers.

These two special areas of robotics require different areas of study. However, people working in robotics will need to know about other areas that relate. Someone who specializes in computer software, for example, will have to understand the mechanics of arm movements in order to program them.

To work with the "brain" or computer mind of the robot, you will need training in computer science. You will study theories on how computers work and you will learn how to develop software. You will also need to study *linguistics*—the science of language. This will be essential, as we have mentioned in the chapter on the robot's "brain," in order to make your robot understand and respond to commands. You will also need to have good grammar and sentence structure and be able to write in a clean, concise manner. Mathematics will be very important—geometry, algebra, and more advanced fields such as trigonometry and calculus. You will need to know physics—how will your robot respond to the physical world around it? For example, how will it react to extreme cold or heat? Will it work in space where there is no gravity?

Other aspects of study you may not have considered but which will be important are *cognitive psychology* (how humans think), *ethics* (the study of what humans consider right and wrong), and *sociology* (the study of how humans act as a society and how they interact with machines).

As we have explained, you will need to know computer science to work on the body. But in addition, you will need some knowledge of mechanical and electrical engineering. You will also need to study human *physiology*—the science of the human body. It helps to know how the human hand works, for example, to be able to design a robot hand to perform the same task. You will also need to know *material science*—the study of materials such as plastic, metal, ceramics and glass, and new materials that haven't even been developed yet.

As you can see, a college degree is essential for work in the field of robotics, although you can have fun, as many young people do now, building "hobby" robots if you have a limited knowledge of mechanical and electrical engineering and computer science.

Above all, however, *imagination* is the key to the future of robotics. It will be people with imagination who will provide answers to the questions we have asked. It will be such people who develop the new technology necessary to build complex, thinking robots.

[47]

# ACTIVITIES

## THE HEADACHES OF BEING A ROBOT

The purpose of this activity is to demonstrate (1) how to write a software program without a computer, (2) how even the simplest task a robot could perform is a complex series of movements, and (3) how precise you must be in giving directions to a robot.

This activity requires at least two people, and can also be done by a group or in a classroom.

### Materials Needed

One pair of heavy mittens or cooking mitts, a blindfold, two wooden sticks and a whistle, pencil and paper, and an object for the robot to pick up such as a large block (nothing breakable!).

One person is chosen to be the "body" of the robot. This person should have a good memory because he or she will need to memorize a series of simple commands. Blindfold this person and put the heavy mittens over the person's hands. This person is a very simple robot who must react only to commands given in the code that has been developed. The "body" may not talk to the "brain." The "brain" may not touch or speak to the body except through the code.

Another person, or several persons, are chosen to be the "brain," or the computer software that is the robot's "brain."

[48]

Since the robot's "brain" can speak to the body only through electronic impluses (described in Chapter 3), a code must be developed to act as these impulses. This can be done in a variety of ways, using musical tones, for example. But we suggest the easiest way is by tapping two wooden sticks together for commands and blowing a whistle to indicate the command has ended and a new one is about to be given. The table on page 51 shows how a code can be developed.

## Command Sequence

Part of the body, direction, how far, end of command

## Writing the Program

In this sample program we are going to instruct the "body" to move forward and lift an object off the table. You can write your own programs using the codes given on page 51, which are fairly simple to memorize. This could even include such complicated commands as going up or down stairs, although that will take a lot of thought! The program should be written first, before being given to the "body," just as a computer programmer writes a program. This will make it easier to give instructions to the "body" and also allow you to see and correct any mistakes.

Position the robot about four feet (1.2 m) away from the table, facing the table, arms at its sides, feet together. The robot is blindfolded with heavy mitts on its hands and must react only to commands given it in code. (You might want to equip your robot with a beeping sound if it doesn't understand a command and wants it repeated.)

## Sample Program

You are going to command the robot to move forward and pick up the object from the table. It is important that you follow the command sequence every time!

**PART OF
THE BODY**

**DIRECTION**

**HOW FAR**

1 tap —right arm
2 taps—left arm
3 taps—right leg
4 taps—left leg

1 tap —up
2 taps—down
3 taps—in
4 taps—out
5 taps—forward
6 taps—backward

Number of taps indicates how far
to move in inches or feet (centimeters or meters) for arm movements and in feet for leg movements

1 short whistle—end of command, beginning of another command*

*It is important that the "body" not react to a new command unless
it hears the whistle! If the "brain" forgets to whistle, the body
must simply keep repeating the last command it has heard!

1. Whistle. Indicates command is coming.
2. 3 taps (right leg)
   pause
   5 taps (forward)
   pause
   1 tap (move forward one foot, or .3 m)
3. Whistle. End of command. New command coming. *(Note: if the whistle is forgotten, the robot will continue moving the right leg forward according to the number of taps given until it falls over!)*
4. 4 taps (left leg)
   pause
   5 taps (forward)
   pause
   2 taps (move forward two feet, or .6 m)
5. Whistle. End of command. New command coming.
6. 3 taps (right leg)
   pause
   5 taps (forward)
   pause
   1 tap (move forward one foot, or .3 m)

Both legs should be together at this point. Repeat same command sequence if robot needs to move closer to the table. Be certain to write entire sequence in your program.

7. Whistle. End of command. New command coming.
8. 1 tap (right arm)
   pause
   1 tap (up)
   pause
   10 taps (move up ten inches, or 25 cm, or however many needed to be level with object)

9. Whistle
10. 2 taps (left arm)
   pause
   1 tap (up)
   10 taps (move up ten inches, or 25 cm, or corresponding
     to above figure)
11. Whistle
12. 1 tap (right arm)
   pause
   3 taps (move right arm in)
   pause
   5 taps (five inches, or 13 cm, or figure needed to have
     hand touch object)
13. Whistle
14. 2 taps (left arm)
   pause
   3 taps (in)
   pause
   5 taps (move in five inches, or 13 cm)

Both robot's "hands" should be firmly "holding" object at this point. Since we do not have a command for both hands to lift at the same time (you might want to add one, although it is not necessary), we will instruct the robot to move backward while holding the object.

15. 3 taps (right leg)
   pause
   6 taps (backward)
   1 tap (move back one foot)
16. Whistle
17. 4 taps (left leg)
   pause

6 taps (backward)

pause

1 tap (move back one foot)

**18.** Whistle

This ends the program.

It would be interesting for you to devise your own system of codes to make the robot operate in more complicated ways. You could devise codes for turning the robot to the left or the right, for raising both hands together, or for climbing stairs. Always remember, though, that you must give the code in the *correct sequence* every time (which leg, what direction, how far). Imagine what would happen if you commanded the robot to climb a stair and forgot to tell it to lift its leg first!

Take turns being the robot's "brain" and "body." This will give you an idea of how each part operates and how they all must work together. Two of you might want to demonstrate to the rest of the class what would happen if you mixed up the sequence of the code commands or if you forgot to blow the whistle. You might even want to write a play starring your "robot"!

# GLOSSARY

*Android:* an artificial being created out of human parts or parts made to look human.

*Arc welding:* a system that uses powerful electric currents to melt metal for the purpose of joining pieces of metal together.

*Artificial intelligence:* the development or capability of a machine to perform functions that are normally concerned with human intelligence, such as learning, adapting, reasoning, self-correction, and automatic improvements.

*Auditory sensor:* a device that turns sounds into electronic signals.

*Depth perception:* the ability to perceive or "see" objects in relation to one another, such as from front to back, from the top straight down, and from the surface or edge inward.

*Industrial robots:* robots used in manufacturing and production.

*Mechanical manipulation:* the ability of a machine to move and handle objects.

*Negative feedback:* a process by which part of the output signal of an amplifying circuit is fed back to the input circuit.

*Optical sensor:* a device that turns visual impulses into electronic signals.

*Pattern recognition:* the ability to see or recognize an arrangement or design.

*Pressure sensor:* a device that can sense a force being exerted against it.

*Program:* a set of instructions arranged in correct sequences to direct a digital computer in performing a desired operation or operations.

*Robot:* a machine created out of mechanical parts.

*Robotics:* the use or study of robots.

*Software:* programs, languages, and procedures of a computer system.

*Sonar:* a device that transmits high-frequency sound waves and registers the vibrations reflected back from an object.

*Stepper motor:* a motor that rotates in a series of steps.

# BIBLIOGRAPHY

Adams, Douglas, *The Hitchhiker's Guide to the Galaxy*, New York: Pocket Books, 1979.

Adams, Douglas, *The Restaurant at the End of the Universe*, New York: Pocket Books, 1980.

Asimov, Isaac, "Robbie," *Masterpieces of Science Fiction*, New York: Ariel Books/Ballantine Books, 1978.

Bongard, M., *Pattern Recognition*, New York: Spartan Books, 1970.

Clarke, Arthur, *2001: A Space Odyssey*, New York: New American Library, 1972.

Cohen, John, *Human Robots in Myth and Science*, New York: A. S. Barnes and Co., 1967.

Elrick, George S., *Science Fiction Handbook*, Chicago: Chicago Review Press, 1978.

Findler, N. V., and Bernard Meltzer, *Artificial Intelligence and Heuristic Programming*, New York: American Elsevier Publishing Company, 1971.

Fogel, Lawrence, Alvin Owens, and Michael Walsh, *Artificial Intelligence Through Simulated Evolution*, New York: John Wiley and Sons, 1966.

Lucas, George, *Star Wars*, New York: Ballantine Books, 1976.

Krasnoff, Barbara, *Robots: Reel to Real*, New York: Arco Publishing, 1982.

Sagan, Carl, *Cosmos*, New York: Random House, 1980.

Warrick, Patricia, *The Cybernetic Imagination in Science Fiction*, Cambridge, Mass., MIT Press, 1980.

For young people interested in pursuing the subject of robotics, we recommend the magazine *Robotics Age*, published by Robotics Age Incorporated, Strand Building, 174 Concord Street, Peterborough, New Hampshire 03458.

# INDEX